AMAZON DASH BUTTON

The Ultimate Guide for Complete Beginners On How to Use the Amazon Dash Button in Few Minutes.

I0482075

BY

JAMES R. HALE

COPYRIGHT©2018

COPYRIGHT

James R. Hale

TABLE OF CONTENT

CHAPTER 1

INTRODUCTION

Amazon Dash Buttons are easy, single button unit that can be hung anywhere around the house. When gotten it connected to a WI-FI network and pressed, they give order for a single commonly stock item from Amazon. It buttons cost £4.99 a pop, but you have a £4.99 reduction on your first dash order, in essence making them basically free.

The idea is that you keep them in a strategic way around the house, so to ease the ordering of more toilet paper when you are on the bog, got stock up on coffee pods in the kitchen, and so on. Each button is separately branded with a particular company's livery, showing the device's singular purpose.

This guide will show you how you can use the Amazon Dash Button with step by step process even as a beginner, all you have to do is just follow the steps as instructed. But if you find this very difficult, there is no needs to worry because I got you cover.

All over the world millions of people haven't been able to use and know the importance of Amazon Dash Button, but this book gives the breakdown of all solution to any problem you might encounter. With proper medication and better understanding you will archive all about Amazon Dash Button in a few minutes

Thankfully each steps are very easy and simple to follow, that even a beginner can master it in a few minutes.

CHAPTER 2

AMAZON DASH BUTTON

Amazon has unveiled its recent UK Dash Button expansion, joining an additional 39 household staples to the one push ordering scheme. Useful additions include cleaning brands Bold, Cillit Bang, and Calgon; pet food mainstays Purina, Bakers, and Felix; and beverage companies The ebglish Tea Shop, Nescafe, and Evian.

Other big name brands having their own Dash Buttons may include Clairol, Pampers, Head and Shoulder, Tampax, and Gaviscon.

The proclamation means that Amazon UK offers Dash Button for more than 100 leading products. Earlier in the year, the firm introduce the likes of Mentos, Duracell, Glade, Joseph Joseph, Scott, Tassimo, L'or, Regina, Perfect Fit, Muscle, Wellman and Wellwoman to the instant order program.

However, if you are extremely reckless to do one of those science experiment with Coca-Cola and Mentos, and unable to go to the shops, Amazon got you covered.

As with former Dash Buttons, the recently made batch of units cost £4.99 each, but it does come with the same amount in credit for the associated brand. Which is the present state of play with Amazon Dash Buttons. Below you will get all details to what Amazon Dash Buttons are and how they work.

WHAT ARE AMAZON DASH BUTTONS?

At the time Amazon make known its instant-ordering Dash Buttons in 2015, we thought it was just a prank, but it appears they were real indeed. Amazon Dash Buttons vow to change radically how to shop, but exactly how they work and reasons you should care will be all explain in this guide.

Amazon Dash Buttons are easy, single button unit that can be hung anywhere around the house. When gotten it connected to a WI-FI network and pressed, they give order for a single commonly stock item from Amazon. It buttons cost £4.99 a pop, but you have a £4.99 reduction on your first dash order, in essence making them basically free.

The idea is that you keep them in a strategic way around the house, so to ease the ordering of more toilet paper when you are on the bog, got stock up on coffee pods in the kitchen, and so on. Each button is separately branded with a particular company's livery, showing the device's singular purpose.

As earlier said, there were 40 supported brands for Amazon Dash's 2016 UK launch, with the number now increasing to 65. Already Amazon has your details on file, so you don't need to do anything else. You just have to click on the button, and the linked product will be delivered by using your 1-click ordering preferences for shipping address and payment method.

CHAPTER 3

GETTING THE AMAZON DASH BUTTONS SETUP

The Amazon Dash buttons are so easy and simple to set up. After having the latest version of the Amazon shopping app for android (or iOS), you can be up and running in a matter of time.

Now, to setup your Amazon Dash Buttons first you have to have the highly valued

Dash Button, the one for delivery new toilet roll. Which is just a case of opening the Dash Button section of the app, ensuring my home WI-FI connection and then pressing the Dash Button downwards until the blue light glint.

Next is to pair the button, making a neat sci-fi/dial-up-internet-style buzzing sound while finishing the task.

After that, next is to choose the product you wish to associate with the button every time you click to reorder. However, in this case I selected my item, ensure payment and shipping information, then affixed the button. There is sticky or a rung to place on a hook next to my toilet.

Now, after pressing on the button for the first time late on a Friday night, spent my Friday night ordering toilet roll from the internet. A green led light glint to assure the order. Immediately, a notification pop up through the app and an email. Pretty smart.

Safety measures are built in with your Amazon Dash unit programmed to only reply to the first press, unless otherwise stated explicitly. Also you will receive an order alert sent directly to your phone whenever one is raised, hence allow you to terminate any input error resulting from fastidious children.

CHAPTER 4

DASH REPLENISHMENT SERVICE

Amazon Dash Replenishment Service brings things one step more, letting products with lot ability automatically record the need to re-stock, so you are able to shipped your crucial without having to press a button.

Some of the manufacturers that take part include Bosch, Samsung, Whirlpool and Brother. Have you ever been in a more teasing moment than finding your ordered printer ink right on your doorstep at exactly the right time?

These live on Amazon's website and app, surfacing shopping proposals based on items you newly or occasionally order. Just like with physical Dash Buttons, hitting a Virtual Dash Button gives an order with a single tap.

However, they are totally free to use, so go for an ideal jumping off point if you are conspired prospect of Dash Buttons, but not ready to put all in.

Note that to Amazon Prime members Amazon Dash Buttons are entirely available. Tap the link below to acquire a free 30-day Amazon Prime trial and confirm them out for yourself, no strings attached.

CHAPTER 5

THE BEST AMAZON DASH BUTTON FOR THE UK

In June, Amazon introduce 20 recently made Dash Buttons to the UK, making the sum number of brands to 65. Given most brands have some little products from their range available, indicating that UK shoppers can get hundreds of items delivered to their door with just a hit.

Totally, the most serviceable Amazon Dash Buttons will vary but depend on your living condition.

However, it appears cleaning products will be at the top of most folk's list. They are stuff you didn't remember to buy at the supermarket when your attention is diverted by the beer and chocolate isles. So, the likes of Finish, Detol, Andrex, Fairy, Lenor, vanish and Ariel have proven to be famous selections.

Toiletries and beauty products from the likes of Gillette, Wilkinson Sword, Olay, Rimmel, Right Guard, Listerine and Neutrogena will make you looking and smelling nice.

Talking of smelling, if you have a new baby in your house then Johnson's and Huggies baby products will demand consistent replenishment. The previous has 21 products to select from, while the later has only one.

Amazon gave report that beyond the listed cleaning brands, Gillette (32 products), Nescafe Dolche Gusto coffee capsules and Whiskas cat foot at this moment are the most famous.

Of the new breed, our expectations of batteries ate from Duracell (close to the Xbox one) and, of course Heineken is among the most popular.

CHAPTER 6

VERDICT

At first, when I told people about the Dash Buttons, their first reply was "can they be program to order anything?" The answer is no. but wouldn't that be something good? If these were ever-present and could be programmed to perform anything, having some of these in your home would be useful.

To the guitarist who just strung his spare set, or for the home brewer who loves to have an extra yeast on hand always. Socks. Some people need more socks most times, right?

Unluckily, a programmable Dash Button is not the directly future, so who are these things foe in their present form? Haste parents, work-shy bachelors and bachelorettes, widely forgetful humans and slaves to the leading brands who are agreeing to pay whatever the cost for the greatest convenience.

If you fall on one or more of those categories, then Dash Buttons are wonderful. Occupy your house with them at once.

THE END

James R. Hale

James R. Hale

James R. Hale

James R. Hale

James R. Hale